PREFACE

Roughly half of this work was done at the Center for Relativity Theory, the University of Texas at Austin, in connection with my Ph.D. dissertation. The second half was done during my first year as a Research Associate with the Physics Department at Syracuse University, after an enlightening week spent with the relativity group at the University of Pittsburgh.

I would like to thank L. C. Shepley, J. Ehlers, and P. G. Bergmann for their great helpfulness and interest in this project. I am deeply indebted to E. T. Newman, for kindly permitting me to discuss some as-yet-unpublished results concerning Heaven theory and for pointing out an error in an earlier version of this manuscript; to J. N. Goldberg for some cogent criticisms of a portion of the manuscript; and to D. Lerner, for many illuminating discussions. I am lucky enough to owe words of thanks to all of the following people: C. DeWitt, D. Cox, D. P. Duncan, R. O. Hansen, A. E. Hwang, L. P. Hughston, R. A. Matzner, R. Parsons, R. Penrose, J. Reed, I. Robinson, A. Schild, J. Vick, J. Weinberg, and J. Winicour. Finally, I would like to thank the typists of the manuscript, R. Wagoner and R. Newsholme, for their apparently infinite patience.

This research was supported in part by an NSF Traineeship and by NSF Grants GP 34639X-1 and GP 43759X.

TABLE OF CONTENTS

TABLE OF CONTENTS

TABLE OF CONTENTS

CHAPTER I: INTRODUCTION

Knowing that by the chain causality
All separate existences are wed
Into one supreme whole ...

... lend fire to the mind,
And being joined with it in harmony
More mystical than that which binds the stars planetary,

Strike from their several tones one octave chord
Whose cadence being measureless would fly
Through all the circling spheres ...

                    -- Oscar Fingal O'Flahertie Wills Wilde

   We have become accustomed to the very basic role which
complex numbers and holomorphic functions play in quantum
theory, particularly that of elementary particles. It
seems, therefore, that complex numbers are (at least at our
present level of understanding) a very important constituent
of the structure of physical laws. ... complex numbers may
also be very basically involved in defining the nature of
space-time itself.

                    -- Roger Penrose

   The study of complex manifolds can be traced back ultimately to the monu-
mental work of Gauss and Riemann on the theory of curved surfaces; the classi-
cal Riemann surfaces were the first examples of complex manifolds. Riemannian
geometry and the theory of several complex variables developed essentially independently until the early 1930's, when the two concepts were synthesized
by Schouten and van Dantzig, and independently by Kähler. The results are
known today as Hermitian and Kählerian geometry. In the early 1940's,
Hodge's work on algebraic varieties laid the foundation for the study of
the special cases of Kählerian manifolds which are today known as Hodge
manifolds. In 1947, Weil and Ehresmann developed the concept of an almost

complex manifold, and the 1950's saw the development of this idea by
Eckmann, Frölicher, Newlander and Nirenberg, and others. Also at this
time, Chern looked at the classification of complex vector bundles, and
examined what are now known as the Chern classes of these bundles. In the
late 1950's and early 1960's, Kodaira, Spencer, and Kuranishi developed the
theory of deformations of complex manifolds and examined the embedding
problem for complex manifolds.

Perhaps no other area of modern mathematics is as broadly based
as complex manifold theory, interweaving the beauties of geometry and
topology with the subtleties of complex analysis. In spite of this,
complex manifold theory is still not considered to be an indispensible
element of a mathematician's education, and research in this area, though it
is decidedly "alive and well", is carried out by a relatively small number
of enthusiastic workers.

The first marriage of relativity and complex variables was produced
by Einstein himself with the introduction of an imaginary time coordinate
for Minkowski space. Ironically enough, this feature was abandoned by him
when general relativity was developed. (Decades later, though, in one version
of the unified field theory, a complex-valued "Hermitian" metric tensor
was used.)

In a more peripheral sense, complex variables began to appear in
several areas of research in general relativity in the 1960's. Examples
of this are the null tetrad approach of Debney, Kerr, and Schild, the spinor
calculus methods of Penrose, and the spin coefficient formalism developed
by Newman and Penrose. Trautman, and independently Newman, examined the
analytic continuation of Maxwell's equations and the linearized Einstein's
equations onto "complex Minkowski space." One of the most novel uses of

complex variables was Newman's derivation of new metrics from known ones
by "complex coordinate transformations." On a different front, Brans
recently showed that the idea of a complex structure on a vector space is
very pertinent to the bivector $(SO(3,C))$ representation of the Lorentz
group. In the present author's Ph.D. dissertation, the standard Hermitian
and Kählerian geometry of positive definite Riemannian spaces was modified
in such a way as to be applicable to spacetime manifolds with their
indefinite metrics, and a mildly interesting but apparently closed set of
results was obtained.

Although many authors (including this one) had purported to
mathematically rigorize Newman's complex coordinate transformations, the
underlying physics of the situation remained mysterious. However, within
the last several years, Newman and Penrose have obtained some exciting
results which appear not only to shed light on this question, but to suggest
that a fundamental revision of our ideas of space and time and even conceivably
of the foundations of physics may soon be brought about. These developments
are known as twistor theory and H-space or Heaven theory.

The "twistor program", as enunciated by Penrose in the late 1960's,
has as its ultimate goal the elimination of the continuum concept in physics
and its replacement by physical laws which are purely combinational in
nature. Present results seem to indicate, for example, that twistor theory
provides a quantum physics which is free of the divergences of conventional
quantum field theories. More pertinent here is the fact that twistor
theory seems to indicate that complex numbers enter into the description of
spacetime in a fundamental and non-trivial way; thus, for example, asymptotic
twistor theory (which appears to be pointing the way to an S-matrix theory
of gravity) is based on the concept of a Kählerian manifold.

Newman's concept of H-space or Heaven associates a four-complex-dimensional manifold—H-space—with any asymptotically flat spacetime. This "heavenly space" possesses a "complex Riemannian metric" which is Ricci-flat and "left-conformally-flat." The Schwarzschild-to-Kerr complex coordinate transformation appears to be reflected in the Heaven-picture as a displacement of a center-of-mass line in an imaginary direction. Newman and Penrose have shown that flat-space twistors can be thought of as geometrical objects defined in complex Minkowski space, and that asymptotic twistors can similarly be viewed as residing in "Heavenly space." In this way twistor theory and H-space theory, which on the surface appear to be quite unrelated, are seen to be in fact intimately connected.

Another surprising aspect of Heaven theory is the fact that Heavenly metrics are a very special case of (the complexification of) the left-Kählerian complex spacetimes examined by the author. This fact, together with the Kählerian nature of twistor manifolds, suggests that physicists may profit from an examination of the differential-geometric aspects of complex manifold theory.

It is the purpose of these notes, then, to provide for physicists an introduction to the differential geometry of complex manifolds. Neither twistor theory nor Heaven theory is in a final perfected state: important new results are being obtained virtually every few weeks. Because of this, an intentional effort has been made in these notes to include speculative questions, incomplete results, and conjectures as to the directions future work might take. No attempt is made here to present a complete "textbook of twistor theory"; that is out of the scope of the lecture note format (and beyond the author's capabilities). Also, the more advanced mathematical aspects of complex manifold theory, such as sheaf cohomology,

deformation theory and the embedding problem, partial differential equations or complex manifolds, the classification of complex vector bundles, etc., are barely mentioned here: these are really topics for a "second course" in the subject. However, it is felt that a familiarity with the contents of these notes will orient the reader to approach these more advanced questions with some degree of confidence.

The prerequisites for understanding this work are an elementary knowledge of general relativity [48], [29], differential geometry [50], [4], and complex analysis [9], [40].

CHAPTER II: COMPLEX STRUCTURES ON VECTOR SPACES

It is helpful to begin the study of complex manifolds by first considering the simpler matter of a complex structure on a real vector space. In so doing, of course, we have in mind a particular example of such a vector space: The tangent space at a point of a differentiable manifold. Thus many of the results derived in this chapter will find direct application in our later work. (The theory of complex structures and more generally "extended-field structures" on vector spaces is an interesting study in itself, having some bearing on topics related to Schur's Lemma; see Nomizu [39] for details.) Our treatment here will be based largely on the presentations of Nomizu [39] and Chern [8].

Complexification of a Real Vector Space

The complexification of the tangent space at a point of spacetime is a familiar idea in relativity; for example, the null-tetrad formalism makes use of "complexified tangent vectors" $m^a = a^a + ib^a$ and $\bar{m}^a = a^a - ib^a$ in order to obtain a tetrad of null vectors ℓ^a, n^a, m^a, $\bar{m}^a$ at each point of spacetime. We now define precisely the __complexification__ of a real vector space:

II.1: DEFINITION: Given a real vector space V, also written (V, R) to emphasize the field over which V is defined, its complexification is the complex vector space V^c, also denoted (V^c, C), where:

(i) $Z = X + iY \in V^c \Longleftrightarrow X, Y \in V$,

(ii) $(X_1 + iY_1) + (X_2 + iY_2) \equiv (X_1 + X_2) + i(Y_1 + Y_2)$

for all $X_1, X_2, Y_1, Y_2 \in V$,

(iii) $(\alpha + i\beta)(X + iY) \equiv (\alpha X - \beta Y) + i(\beta X + \alpha Y)$ for

all $X, Y \in V$ and $\alpha, \beta \in \mathbb{R}$.

Complex conjugation in V^c is defined by

$\bar{Z} = \overline{X + iY} \equiv X - iY.$

II.2: REMARK: (i)   It is easy to show that V^c satisfies the

axioms (see Nomizu[39]) for a complex vector

space when V satisfies those for a real vector

space.

(ii)   The (complex) dimension of V^c is equal to the

(real) dimension of V.

Complex Structure

The complexification of a real tangent space does not provide
enough structure for the consideration of complex coordinate transformations
on the underlying manifold. With this in mind, we now proceed to introduce
additional structure on vector spaces:

II.3: DEFINITION: A complex structure on a finite-dimensional real

vector space V is an endomorphism J (i.e., $J:V \to V$

linearly) such that $J(J(X)) = -X$ for all $X \in V$, or

$J^2 = -id$, where id denotes the identity endomorphism.

We denote a real vector space with a given complex structure by
$(V, \mathbb{R}; J)$. As a word of caution, we note that the transformation defined by
$J(X) = iX$ is not a complex structure, because it does not map V into V.

II.4: THEOREM: A real vector space $(V, \mathbb{R}; J)$ with a complex

structure J is even-dimensional.

Proof: Let dim V = n and let $\{e_1, \ldots, e_n\}$ be a basis for V. Then,

since J is linear, it is determined by the n^2 _real_ quantities

$J_a{}^b$, where

$$J(e_a) \equiv J_a{}^b e_b$$

(summation convention employed throughout). Then using the

property $J^2 = -id$, we have:

$$-e_a = J^2(e_a) = J_a{}^b J_b{}^c e_c \Rightarrow J_a{}^b J_b{}^c = -\delta_a{}^c$$

Considering this as a matrix equation, we can write

$$\det(J^2) = (\det J)^2 = \det(-I_n) = (-1)^n$$

where I_n denotes the nxn unit matrix. Thus

$$(\det J)^2 = (-1)^n \geqslant 0 \Rightarrow n = 2m, \qquad m \text{ an integer.}$$

Because of this theorem, we shall adopt the following conventions
in the sequel: small Latin indices on a quantity range and sum over
1, ..., n = 2m, where n = 2m is the dimension of the vector space
(later, of the manifold) in question; small Greek indices range and sum
over 1, ..., m. Later, we shall also have occasion to let barred Greek
indices $\bar{\alpha}, \bar{\beta}, \ldots$ range and sum over m + 1, ..., m + m = n, that is to
say, $\bar{\alpha} = m + \alpha$. The summation convention shall extend to expressions
such as $A_\alpha B^{\bar{\alpha}} = \sum_{\alpha=1}^{m} A_\alpha B^{m+\alpha}$. To avoid confusion, we will restate these
conventions as often as necessary.

The proof of the above theorem illustrates how a complex structure

"mimics" the action of multiplication of vectors by i (for example, the statement $J_a{}^b J_b{}^c = -\delta_a{}^c$ is suggestive of $i^2 = -1$). In fact, given a vector space $(V, \mathbb{R}; J)$ with a complex structure, we can make the set V into a complex vector space, denoted $(V, \mathbb{C}; J)$, in the following way: the addition in $(V, \mathbb{R})$ is retained and complex scalar multiplication is defined by

$$(\alpha + i\beta)X \equiv \alpha X + \beta J(X) \quad \text{for all } \alpha, \beta \in \mathbb{R} \text{ and } X \in V .$$

II.5: REMARK: (i)   It is easy to show that $(V, \mathbb{C}; J)$ with this addition and multiplication satisfies the axioms for a complex vector space.

(ii)   If $\dim(V, \mathbb{R}) = n = 2m$, then $\dim(V, \mathbb{C}; J) = m$.

The Relation Between Complexification and Complex Structure

The second part of the remark above shows that $(V, \mathbb{C}; J)$ cannot be the same as $(V^c, \mathbb{C})$, since their dimensions are not equal. However, there is a natural relationship between a complex structure on a real vector space and the complexification of that vector space. Given $(V, \mathbb{R}; J)$, we define the following subsets of V^c:

$$W(J) \equiv \{Z \mid Z = X - iJ(X) \text{ and } X \in V\} ,$$

$$\overline{W}(J) \equiv \{Z \mid Z = X + iJ(X) \text{ and } X \in V\} .$$

II.6: REMARK: (i) $W(J)$ and $\overline{W}(J)$ are subspaces of the complexification V^c of V.

(ii) $W(J)$ and $\overline{W}(J)$ are complex conjugate to

each other, in the sense that $Z \in W(J) \Longleftrightarrow$ $\bar{Z} \in \bar{W}(J)$ for all $Z \in V^C$.

Then we have the following result:

II.7: THEOREM: If $(V, R; J)$ is a real vector space with a complex structure, then $V^C = W(J) \oplus \bar{W}(J)$ (direct sum). Conversely, given (V, R), U and $\bar{U}$ such that U and $\bar{U}$ are subspaces of (V^C, C) which are complex conjugate to each other and $V^C = U \oplus \bar{U}$, then there exists a complex structure J for V such that $W(J) = U$ and $\bar{W}(J) = \bar{U}$.

Proof: To prove that $V^C = W(J) \oplus \bar{W}(J)$, we must show:

(i) $W(J) \cap \bar{W}(J) = \{0\}$ and

(ii) if $Z \in V^C$ then $Z = Y_1 + Y_2$ with $Y_1 \in W(J)$ and $Y_2 \in \bar{W}(J)$.

To establish (i), consider $X_1 - iJ(X_1) \in W(J)$ and $X_2 + iJ(X_2) \in \bar{W}(J)$ where $X_1, X_2 \in V$. If

$$X_1 - iJ(X_1) = X_2 + iJ(X_2) ,$$

then $X_1 - X_2 = J(X_1 + X_2) = 0 \Rightarrow X_1 = X_2 = 0$. As for (ii), if $Z = X_1 + iX_2 \in V^C$ where $X_1, X_2 \in V$, then $Z = Y_1 + Y_2$ where

$$2Y_1 = X_1 + J(X_2) - iJ(X_1 + J(X_2)) \in W(J)$$

and $2Y_2 = X_1 - J(X_2) + iJ(X_1 - J(X_2)) \in \bar{W}(J)$.

To prove the second half of the theorem, suppose $V^C = U \oplus \bar{U}$ with U and $\bar{U}$ complex conjugate subspaces of V^C. Now V may be identified with the subset of V^C consisting of all vectors $X \in V^C$

such that $X = \bar{X}$. Any such vector X may be written uniquely as $X = \frac{1}{2}(Z_1 + \bar{Z}_2)$ where Z_1, $Z_2 \in U$, and

$$X = \bar{X} \Rightarrow Z_1 + \bar{Z}_2 = Z_2 + \bar{Z}_1 \Rightarrow Z_1 = Z_2 .$$

We can then define an endomorphism $J:V \to V$ by

$$J(X) = \frac{1}{2}(Z_1 - \bar{Z}_1) \equiv Y$$

where $Y = \bar{Y} \in V \subseteq V^c$. Then J is a complex structure for V since $J^2(X) = -X$ for all $X \in V$. Furthermore, $W(J) \subseteq U$ since for any $X \in V$,

$$X - iJ(X) = \frac{1}{2}(Z_1 + \bar{Z}_1) - i\frac{1}{2}(Z_1 - \bar{Z}_1) = Z_1 \in U .$$

Then since any $Z_1 \in U$ defines a member of $W(J)$, we have $U \subseteq W(J)$ and thus $U = W(J)$. By a similar argument, $\bar{U} = \bar{W}(J)$.

This theorem shows that a complex structure J on V is equivalent to a choice of complex conjugate subspaces $W(J)$ and $\bar{W}(J)$ of the complexification V^c with the property $V^c = W(J) \oplus \bar{W}(J)$.

II.8: REMARK: The endomorphism $J:V \to V$ can be extended to an endomorphism $J:V^c \to V^c$ by defining $J(X + iY)$ $J(X) + iJ(Y)$. Then $W(J)$ is that subspace of V^c consisting of all vectors Z such that $J(Z) = iZ$. Similarly, $\bar{W}(J)$ is uniquely defined as all vectors $\bar{Z}$ such that $J(\bar{Z}) = -i\bar{Z}$.

Given the vector space V and the complex structure J associated with a direct sum decomposition $V^c = W(J) \oplus \bar{W}(J)$ of the complexification, then vectors $Z \in W(J)$ are said to be of type (1,0) and vectors $\bar{Z} \in \bar{W}(J)$

are said to be of <u>type (0,1)</u>.

Conjugate Complex Structure

If J is a complex structure for V, then clearly the endomorphism $-J:V \to V$ is also a complex structure; $-J$ is said to be the complex structure <u>conjugate</u> to J.

II.9: <u>REMARK</u>:   The vectors of type $(1,0)$ (type $(0,1)$) with respect to J in V^C are the vectors of type $(0,1)$ (type $(1,0)$) with respect to the conjugate complex structure $-J$.

Complexification of the Dual Space

We now consider the dual space V^* of $(V, \mathbb{R}; J)$, consisting of all linear functions $\omega:V \to \mathbb{R}$. The value of $\omega \in V^*$ at $X \in V$ will be denoted by either $\omega(X)$ or $\langle X, \omega \rangle$. Let V^{*C} denote the complexification of V^*; then V^{*C} may be identified with the complex vector space of all linear functions $\omega + i\theta:V \to \mathbb{C}$ if we set $(\omega + i\theta)(X) = \omega(X) + i\theta(X)$ for all $X \in V$ and for all $\omega + i\theta \in V^{*C}$.

II.10: <u>REMARK</u>:   Defining V^{C*}, the dual of V^C, there is an obvious one-to-one and onto mapping from V^{C*} to V^{*C} given by $\zeta \to \xi + i\eta$, where $\zeta \in V^{C*}$ and $\xi, \eta \in V^*$.

The complex structure J on V induces a complex structure J^* on V^* in a natural (and unique) way. For any $\omega \in V^*$ we define $J^*(\omega) \in V^*$ by

$$J*(\omega)(X) = \omega\big(J(X)\big) \quad \text{for all } X \in V \ .$$

Then $J*$ is a complex structure on $V*$ because

$$(J*)^2 (\omega)(X) = J*(\omega) \big(J(X)\big)$$
$$= \omega\big(J^2(X)\big)$$
$$= \omega(-X) = (-\omega)(X) \ .$$
$$\Rightarrow \quad (J*)^2(\omega) = -\omega \quad \text{for all } \omega \in V* \ .$$

By Remark II.8 above, the elements of type $(1,0)$ in $V*^C$ are those functions $\omega: V \to \mathbb{C}$ for which $J*(\omega) = i\omega$, which is to say

$$\omega\big(J(X)\big) = i\omega(X) \quad \text{for all } X \in V \ .$$

Similarly the elements of type $(0,1)$ are those $\bar{\omega}$ for which

$$\bar{\omega}\big(J(X)\big) = -i\bar{\omega}(X) \quad \text{for all } X \in V \ .$$

From what we have done above, we see that this designation of vectors in $V*^C$ provides a direct sum decomposition into complex conjugate subspaces, $V*^C = W(J*) \oplus \bar{W}(J*)$.

Expressions in Terms of Bases

Let $\{e_1, \ldots, e_n\}$ be a basis for V and let $\{e*^1, \ldots, e*^n\}$ be the dual basis of $V*$, that is to say,

$$e*^a(e_b) = \delta_a{}^b \ .$$

Then any element of V can be written as $X = x^a e_a$ and any element of $V*$ can be written as $\omega = \omega_a e*^a$ (recall that Latin indices range and sum over $1, \ldots, n = 2m$).

In the proof of Theorem II.4 it was noted that a complex structure J can be characterized by the $n \times n$ matrix $J_a{}^b$, where

$$J(e_a) = J_a{}^b e_b \quad ; \quad J_a{}^b \in \mathbb{R}.$$

It follows that if $X = x^a e_a$ and $J(X) = Y = y^b e_b$, then

$$y^b = J_a{}^b x^a \quad .$$

Similarly, if $\omega = \omega_a e^{*a}$ and $J^*(\omega) = \theta = \theta_b e^{*b}$, then

$$\theta_b = J_b{}^a \omega_a \quad .$$

It was also noted in the proof of Theorem II.4 that $J_a{}^b J_b{}^c = -\delta_a{}^c$. We can use this fact to find the eigenvalues of $J_a{}^b$; if $J_a{}^b z^a = \lambda z^b$, then we have:

$$J_b{}^c J_a{}^b z^a = -z^c = \lambda J_b{}^c z^b$$
$$\Rightarrow \quad -z^c = \lambda^2 z^c$$
$$\Rightarrow \quad \lambda = \pm i \quad .$$

Since $J_a{}^b$ is a real $2m \times 2m$ matrix, it follows that $J_a{}^b$ has m eigenvalues $+i$ and m eigenvalues $-i$. Notice that the (complex) eigenvectors corresponding to the eigenvalue $+i$ are vectors of type $(1,0)$ and those corresponding to the eigenvalue $-i$ are vectors of type $(0,1)$.

II.11: REMARK: The most general complex structure on a vector space is described by a matrix $J_a{}^d$ with respect to an arbitrary basis $\{e_1, \ldots, e_n\}$, where

$$J_a{}^d = S_a{}^b (J_0)_b{}^c (S^{-1})_c{}^d \quad ,$$

and $(J_0)_b{}^c = \begin{pmatrix} 0 & I_m \\ -I_m & 0 \end{pmatrix}$,

where I_m is the $m \times m$ unit matrix, and $S_a{}^b$ is a real non-singular matrix.

Given the basis $\{e*^1, \ldots, e*^n\}$ for $V*$, we define elements λ^a of $V*^c$ by use of J:

$$\lambda^a(X) \equiv e*^a(J(X)) + ie*^a(X) = \langle J(X), e*^a \rangle + i\langle X, e*^a \rangle .$$

II.12: REMARK: $\lambda^a \in V*^c$ is of type $(1,0)$ and its complex conjugate, defined by

$$\overline{\lambda}^a(X) = \langle J(X), e*^a \rangle - i\langle X, e*^a \rangle ,$$

is of type $(0,1)$.

In terms of the bases given above, we have:

$$\begin{aligned}
\lambda^a(X) &= x^b \langle J(e_b), e*^a \rangle + i\, x^b \langle e_b, e*^a \rangle \\
&= x^b J_b{}^c \langle e_c, e*^a \rangle + i\, x^b \delta_b{}^c \langle e_c, e*^a \rangle \\
&= x^b(J_b{}^a + i\, \delta_b{}^a) .
\end{aligned}$$

Since J has exactly m eigenvalues $-i$, it follows that the rank of the $2m \times 2m$ matrix $(J_b{}^a + i\, \delta_b{}^a)$ is m. Therefore exactly m of the $n = 2m$ functions λ^a are $\mathbb{C}$-linearly independent. Suppose the basis $\{e*^a\}$ is chosen in such a way that $\lambda^1, \ldots, \lambda^m$ are $\mathbb{C}$-linearly independent. Then clearly $\lambda^1, \ldots, \lambda^m$ form a basis for the subspace $W(J*)$ of $V*^c$ (and $\overline{\lambda}^1, \ldots, \overline{\lambda}^m$ form a basis for $\overline{W}(J*)$). We write such λ^α in terms of real and imaginary parts as

$$\lambda^{\alpha} = \mu^{\alpha} + i\mu^{\bar{\alpha}}$$

(recall that $\alpha, \beta, \ldots$ range and sum over $1, \ldots, m$ and $\bar{\alpha}, \bar{\beta}, \ldots$ over $m+1, \ldots, m+m = n$; note in particular that $\mu^{\bar{\alpha}}$ is a quantity different from μ^{β}).

II.13: __THEOREM:__ The elements $\mu^1, \mu^{\bar{1}}, \ldots, \mu^m, \mu^{\bar{m}}$ are R-linearly independent.

__Proof:__ Suppose $B_{\alpha}\,\mu^{\alpha}(X) + C_{\bar{\alpha}}\,\mu^{\bar{\alpha}}(X) = 0$ for all $X \in V$, with $B_{\alpha}, C_{\bar{\alpha}} \in \mathbb{R}$. Then using $\mu^{\alpha} = \frac{1}{2}(\lambda^{\alpha} + \bar{\lambda}^{\alpha})$ and $\mu^{\bar{\alpha}} = \frac{1}{2i}(\lambda^{\alpha} - \bar{\lambda}^{\alpha})$ we obtain:

$$(B_{\alpha} - i\,C_{\bar{\alpha}})\lambda^{\alpha}(X) + (B_{\alpha} + i\,C_{\bar{\alpha}})\bar{\lambda}^{\alpha}(X) = 0 .$$

(Recall that $C_{\bar{\alpha}}\lambda^{\alpha}$ is a summation, $C_{\bar{\alpha}}\lambda^{\alpha} = \sum\limits_{\alpha=1}^{m} C_{\alpha+m}\lambda^{\alpha}$.)

Replacing X with $J(X)$ and using the fact that λ^{α} is of type $(1,0)$ and $\bar{\lambda}^{\alpha}$ of type $(0,1)$ we can also write:

$$(B_{\alpha} - i\,C_{\bar{\alpha}})\lambda^{\alpha}(X) - (B_{\alpha} + i\,C_{\bar{\alpha}})\bar{\lambda}^{\alpha}(X) = 0 .$$

Together these two equations imply

$$(B_{\alpha} - i\,C_{\bar{\alpha}})\lambda^{\alpha}(X) = 0 \qquad \text{for all } X \in V.$$

But the λ^{α} were chosen to be C-linearly independent. Therefore $B_{\alpha} - i\,C_{\bar{\alpha}} = 0$ for all $\alpha, \bar{\alpha}$, and consequently $B_{\alpha} = C_{\bar{\alpha}} = 0$ so that the $\mu^{\alpha}, \mu^{\bar{\alpha}}$ are $\mathbb{R}$-linearly independent.

This theorem establishes the fact that $\{\mu^{\alpha}, \mu^{\bar{\alpha}}\}$ is a basis for V^* (the μ^{α} and $\mu^{\bar{\alpha}}$ are $\mathbb{R}$-linearly independent and span V^*, since there are $2m = n = \dim V^*$ of them). Let $E_{\alpha}, E_{\bar{\alpha}}$ (where $1 \leq \alpha \leq m$ and $m + 1 \leq \bar{\alpha} = m + \alpha \leq 2m$) be the dual basis of V, so that

$$\mu^{\alpha}(E_{\beta}) = \delta_{\beta}^{\alpha} \qquad \mu^{\bar{a}}(E_{\beta}) = 0$$

$$\mu^{\alpha}(E_{\bar{\beta}}) = 0 \qquad \mu^{\bar{a}}(E_{\bar{\beta}}) = \delta_{\bar{\beta}}^{\bar{a}} \; .$$

Then we can write

$$\lambda^{\alpha}(E_{\beta}) = \mu^{\alpha}(E_{\beta}) + i\mu^{\bar{a}}(E_{\beta}) = \delta_{\beta}^{\alpha}$$

$$\lambda^{\alpha}(E_{\bar{\beta}}) = \mu^{\alpha}(E_{\bar{\beta}}) + i\mu^{\bar{a}}(E_{\bar{\beta}}) = i\delta_{\bar{\beta}}^{\bar{a}} \tag{i}$$

Since λ^{α} is of type (1,0), we also have:

$$\lambda^{\alpha}(J(X)) = i\lambda^{\alpha}(X) = -\mu^{\bar{a}}(X) + i\mu^{\alpha}(X)$$

$$\Rightarrow \quad \lambda^{\alpha}(J(E_{\beta})) = -\mu^{\bar{a}}(E_{\beta}) + i\mu^{\alpha}(E_{\beta}) = i\delta_{\beta}^{\alpha} \tag{ii}$$

$$\lambda^{\alpha}(J(E_{\bar{\beta}})) = -\mu^{\bar{a}}(E_{\bar{\beta}}) + i\mu^{\alpha}(E_{\bar{\beta}}) = -\delta_{\bar{\beta}}^{\bar{a}}$$

From (i) and (ii) we conclude that:

$$J(E_{\beta}) = E_{\bar{\beta}}$$

$$J(E_{\bar{\beta}}) = -E_{\beta}$$

II.14: REMARK: It is easy to show that

$$\lambda^{\alpha}(X) = x^{\alpha} + i x^{\alpha+m}$$

and $$\bar{\lambda}^{\alpha}(X) = x^{\alpha} - i x^{\alpha+m} \; ,$$

where $$X = x^{\alpha} E_{\alpha} + x^{\alpha+m} E_{\bar{\alpha}} \; .$$

Orientations

Recall that an orientation for V is an equivalence class of ordered bases $\{[E_1, \ldots, E_n], [E_1', \ldots, E_n'], \ldots\}$ such that if $E^{(i)}_a = A_a^{\; b} E^{(j)}_b$ for any $[E^{(i)}_a]$ and $[E^{(j)}_b]$ in the class, then $\det(A_a^{\; b}) > 0$. We shall

show that a complex structure defines an orientation on a vector space, but first we need to establish the following result:

II.15: LEMMA: $\det \begin{pmatrix} A & B \\ -B & A \end{pmatrix} = |\det(A + iB)|^2$, where A and B are real m×m matrices.

Proof:
$$\det \begin{pmatrix} A & B \\ -B & A \end{pmatrix} = \det \begin{pmatrix} A + iB & B \\ -B + iA & A \end{pmatrix}$$

$$= \det \begin{pmatrix} A + iB & B \\ -B + iA - i(A + iB) & A - iB \end{pmatrix}$$

$$= \det(A + iB) \det(A - iB)$$

$$= \det(A + iB) \overline{\det(A + iB)}$$

$$= |\det(A + iB)|^2 \quad ,$$

as desired.

II.16: THEOREM: A complex structure J determines an orientation of V.

Proof: It was noted above that the λ^α form a basis for $W(J^*)$. A change in basis of $W(J^*)$ will be given by

$$\lambda'^\alpha = (A_\beta{}^\alpha + i\, B_\beta{}^\alpha)\mu^\beta + i(A_{\bar\beta}{}^{\bar\alpha} + i\, B_{\bar\beta}{}^{\bar\alpha})\mu^{\bar\beta} \quad ,$$

where $A_\beta{}^\alpha$ and $B_\beta{}^\alpha$ are real and $A_{\bar\beta}{}^{\bar\alpha} = A_\beta{}^\alpha$ (i.e., the numerical values of the entries $A_\beta{}^\alpha$ and $A_{\bar\beta}{}^{\bar\alpha} = A_{m+\beta}{}^{m+\alpha}$ are the same) and similarly $B_{\bar\beta}{}^{\bar\alpha} = B_\beta{}^\alpha$. This simply corresponds to a different choice of functions λ^α consistent with the given complex structure J. Then we have:

$$\mu'^\alpha = \mathrm{Re}\,\lambda'^\alpha = A_\beta{}^\alpha \mu^\beta - B_{\bar\beta}{}^{\bar\alpha} \mu^{\bar\beta}$$

Author
Dr. Edward J. Flaherty
Dept. of Physics
Syracuse University
Syracuse, N. Y. 13210/USA

ISBN 978-3-540-07540-0   ISBN 978-3-540-38068-9 (eBook)
DOI 10.1007/978-3-540-38068-9

Lecture Notes in Physics

Edited by J. Ehlers, München, K. Hepp, Zürich,
H. A. Weidenmüller, Heidelberg, and J. Zittartz, Köln
Managing Editor: W. Beiglböck, Heidelberg

46

E. J. Flaherty

Hermitian and Kählerian Geometry in Relativity

Springer-Verlag
Berlin Heidelberg GmbH 1976